YOUR KNOWLEDGE HAS VALUE

- We will publish your bachelor's and
  master's thesis, essays and papers

- Your own eBook and book -
  sold worldwide in all relevant shops

- Earn money with each sale

Upload your text at www.GRIN.com
and publish for free

Bibliographic information published by the German National Library:

The German National Library lists this publication in the National Bibliography;
detailed bibliographic data are available on the Internet at http://dnb.dnb.de .

Imprint:

Copyright © 2009 GRIN Verlag, Open Publishing GmbH
Print and binding: Books on Demand GmbH, Norderstedt Germany
ISBN: 9783656989288

This book at GRIN:

http://www.grin.com/en/e-book/337000/can-people-plantation-forest-policy-stimu-
late-independent-community-based

Omar Pidani, Fitri Nurfatriani

Can 'People Plantation Forest' policy stimulate independent community-based tree growing activities in Indonesia?

GRIN Publishing

Can 'People Plantation Forest' policy stimulate independent community-based tree growing activities in Indonesia?

Abstract

Forest plantations are important in Indonesia for both conservation and development aspects of forest management. They can provide a sustainable supply of wood resources to meet the increasing demands of wood processing industries, rather than escalating pressure on natural forests reserved for conservation. Income from plantation forests can address the economic marginalisation of forest dependent people. Over the last three decades, three strategies have been put into practice to stimulate the development of both large-scale and small-scale plantation forestry in Indonesia: farm forestry, community forestry and community-company partnership. The success, however, has been limited. This paper reviews experience of these strategies in Indonesia, and considers this in the context of criteria and indicators for sustainable plantation development suggested in the literature. It then develops an analytical framework to assess whether a new policy proposed in Indonesia, "the People Plantation Forest" (PPF) policy, is likely to stimulate community-based tree growing activities.

Our analysis suggests that out of six elements identified in the framework, local institutional and capacity building, along with production technology and market access improvement are aspects that PPF might cope well and thus likely to encourage independent community-based tree growing activities. Whereas other elements such land and crop tenure security together with complex licensing and marketing bureaucracy are not dealt with thoroughly and consequently might still be major stumbling blocks in that regard.

For PPF to stimulate independent community-based tree growing, it requires commitment of government agencies across different jurisdictions to coordinate on the provision of technical, financial and regulatory support to minimise constraints in tree growing. Tenure security issue can be minimised through a more participatory approach for land demarcation and mapping; any initiatives conducted by members of community for such purpose should be accommodated. While complex licensing and marketing bureaucracy might be eased off through the creation of a simpler and more integrated procedure. This commitment of support, though, needs to be maintained in the long run given the nature of tree-growing ventures.

Keywords: smallholder tree growing, *Hutan Tanaman Rakyat*

---

[1] College of Asia and the Pacific, Australian National University, ACT, 0200, Australia

[2] Research Center for Forest Socio-Economic and Policy, Indonesian Forestry Department, Jalan Gunung Batu No.5, Bogor, West-Java

Contents

I. Introduction

In Indonesia, community-based tree growing activities have been practiced over a long period of time. Amongst rural and forest margin people, such activities are mainly for subsistence production and income generation. However, the profit that the farmers obtain from these has been a little compare to other commercial timber production activities (Nawir et al., 2007). Underpinning this is a set of socio-economic and ecological issues. Minimum profit that farmers reap also explains why the poverty in and around forest areas is still high (Pradhan et al., 2000).

The problem of rural poverty, along with others such as unsustainable supply and high demand of wood from production forests, minimum access to state forest lands, and maintenance of a wood production system have driven the government to adopt community-based tree growing approaches into its strategies. There have been a range of programs being put in place to address forest-related problems as above. Some of these programs are: farm forestry (*Hutan Rakyat*), community forestry (*Hutan Kemasyarakatan*) and company-community partnership (*kemitraan perusahaan-masyarakat*). Each of these strategies offers both advantages and disadvantages. Hence, their impacts on the ground have been variable (Kusumanto and Sirait, 2000; Poffenberg, 2006).

In 2007, the government issued a new decree regarding the 'People Plantation Policy' (*Hutan Tanaman Rakyat*). This policy is aimed at enhancing the contribution of forestry sector in the development through the improvement of production forest's quality, the alleviation of poverty level and the improvement of the rural and forest-margin people's livelihood (Ministry of Forestry, 2006). There are not many information as to whether or not this policy will be able to remove the key impediments to community-based tree growing activities. Nawir and ComForLink (2007) look specifically into 'partnership' and 'developer' model and argue that this policy fails to provide a proper guideline for partnership. Nordwijk et al., (2007) argues that this policy is not yet able to cope with key constraints of the past strategies such as tenure status, quality wood production capacity of the local people, over arching regulation to market access, and lack of reward for environmental services, yet they have not touched comprehensively on other aspects such as crop protection capacities, licensing procedure, and institutional and capacity building.

Therefore, this paper contributes to the on-going debate about the extent to which the "People Plantation Forest' (PPF) policy will be able to encourage farmers to further integrate trees into their farming, particularly through its independent model (will be discussed in section IV). The paper first reviews benefits and constraints of the past policies of community-based tree growing in Indonesia and combine them with other criteria and indicators for sustainable small-scale plantation as suggested in literatures, particularly by Byron (2001) and Nordwijk et al. (2007), in developing an analytical framework to assess independent scheme of PPF policy.

The rest of the paper is divided into five sections. The next section briefly describes the research methodology applied. After that, it explores the existing strategies of community-based tree growing strategies in Indonesia and their respective key benefits and constraints. Subsequently, it describes the main attributes of the new 'People Plantation Forest' Policy. The final part of this paper includes discussions and policy recommendation.

II. Methodology

2.1. The Research Methodology: the Thematic Analysis

Methodological approach used in this paper is thematic analysis. Thematic analysis is a method for identifying, analysing and reporting themes of patterns within data (Boyatzis, 1998, Joffe and Yardle, 2004, Braun and Clarke, 2006). Unlike its twin, the content analysis, that tends to describe numerically features of particular texts or images, thematic analysis explores more on qualitative aspects of material analysed (Joffe and Yardle, 2004). Thematic analysis used to be applied more in psychological and clinical trial researches but is increasingly used in others policy research topics (for example, see Agar, 1983).

2.2. Data Collection Method and Analysis

The data collection method used for this paper is mainly document study. Documents reviewed were secondary data and included scholarly publication such as peer-reviewed journals and articles, book chapters and the official government regulations and reports. In particular, the review over past strategies of community-based tree growing in Indonesia utilised data published over the period of 1998 until recently. The analysis was conducted manually, and involves reading the documents, taking notes and assessing them against official reports and government documents.

III. The existing community-based tree growing strategies

3.1. Farm Forestry (*Hutan Rakyat* / HR)

Farm Forestry (*Hutan Rakyat*)[3] is defined in Indonesia's *Forestry Act 1991* as tree-growing activities undertaken on community-owned lands. The Farm Forestry program was initially adopted by the government for re-greening[4] and reforestation[5] program in around 1970 (Nawir, 1998) targeting the areas of Java and the Outer Islands (BAPPENAS, 1998; Suharjito et al., 2000 cited in Subarudi et al., 2004). It also has been associated with other socio-economic objectives such as livelihood improvement of rural and forest-margin people, and fulfillment of log demand for wood processing industries (BAPPENAS, 1998; Suharjito et al., 2000 cited in Subarudi et al., 2004). There are approximately 3.43 million of farming households applying a farm forestry approach on their land (FAO, 2001 cited in Nawir et al., 2007). Common species include *Sengon*[6] (*Paraserianthes falcataria*) and *Jati* (*Tectona grandis*) (Nemoto, 2002; Nawir et al., 2007[7]).

To accelerate the implementation of farm forestry (FF) programs, the government has provided various incentives for participating farmers. These incentives range from credit schemes and production input subsidies, which are generally channeled through central government agencies, and technical assistance which is handed over to the local government (as stated in Government Regulation (PP) No.62/1998). For instance, circa 1990, the Directorate General of Reforestation and Land Rehabilitation launched a scheme, the so-called 'farm forestry credit

---

[3] also termed as farm forestry or private forest
[4] Rehabilitation of critical land outside of state domain
[5] Rehabilitation of critical land within state domain
[6] Generally termed as *Sengon* by the local people, although often identified as different species such as *Albizia falcataria* (L.) Fosberg, *Paraserianthes falcataria* (L.) Nielsen, Albizia falcate (L.) Backer, Albizia Moluccana Miq. And Adenanthera falcataria L. (Nemoto, 2002).
[7] Stimulating smallholder tree planting – lessons from Africa and Asia

scheme' (*Kredit Hutan Rakyat* / KHR) under which participating farmers were offered financial incentives (up to Rp. 2 Million per hectare at a subsidized interest rate of 6% per annum) to plant trees on their land for the provision local goods and services. Another similar scheme was the 'credit for watershed conservation' (*Kredit Usaha Konservasi Daerah Aliran Sungai: KUKDAS*) which was aimed at rehabilitation of critical lands within catchments.

3.2. Community Forestry Program (*Program Hutan Kemasyarakatan*)

A Community Forestry (*Hutan Kemasyarakatan*) program is commonly defined in Indonesia as tree-planting activities within state production and protection forests which is aimed at allowing greater access for the community to forest land and resources (Nawir et al., 2007b cited in Nawir, 2008). Until 2003, CF program was regulated by a series of decrees, namely the Ministry of Forestry decrees (SK) No.622/1995, SK No.677/1997, SK No.865/1999, and SK 31/2001. Through these decrees, the community is given a range of rights such as the right to utilize non-timber forest products (SK No.622), the right to manage the forests under 25 year leases; provided community grouped together into a cooperative (SK No.677), the rights to utilize wood and non-wood forest products (SK No.865) and the rights to implement the program (SK No.31). In 2004, a new Ministry of Forestry regulation (*Permenhut*) No.1/Menhut-II/2004 regarding 'Empowerment of people living in and surrounding forest area toward Social Forestry' was issued. Social Forestry (SF) program is slightly different to CF program in terms of rights' recipients and administration complexities. SF program target both private and state forest areas and requires proposer to submit a much more detail proposal which covered the management plan (*kelola Area*), institutional arrangements (*kelola lembaga*), and business arrangements (*kelola bisnis*).

Although the Community Forestry (CF) Program is perhaps the most advanced and comprehensive effort so far to encourage community engagement in forest management (BAPENNAS, 1998), the impact on the ground has been limited (Fay and de Foresta, 1998; Kusumanto and Sirait, 2000; Poffenberg, 2006)(See Figure 2). This can be seen from the percentage of forest area under community-management which is small (0.5%) compared to the total 135 million hectares of forest areas in Indonesia (Poffenberg, 2006). There are various socio-economic and technical reasons for this the slow progress and some of them will be discussed in a latter section of this paper (see Section 4).

3.3. Community-Company Partnership Program (*Program Kemitraan*)

Although 'Community-Company Partnership' (CCP) has been an obligation for timber concessionaires to set up as stipulated in the forestry act and regulations[8], it has become more prevalent since 1999/2000 (Awang et al., 2004; Nawir, 2008). In Java, the CCP program was largely initiated by *Perhutani* (state-owned company) through 'Community Collaboration Management Program' (*Pengelolaan Hutan Bersama Masyarakat* / PHBM), while in the outer islands it was pioneered by several timber-estate companies (Nawir and Gumartini, 2003; Maturana et al., 2005; Awang et al., 2008). In both regions, the development of the CCP program has been driven by the escalating land-related conflict between community and companies especially post-decentralisation period in 1999, and economic marginalisation of forest dwellers (Nawir et al., 2003;Maturana et al., 2005; Awang et al., 2008). Since such conflicts have often been poorly understood by the authorities and create uncertainties for business, some companies have taken the initiative to build partnerships with the communities. This is expected to minimise unexpected risks and at the same time improve the local communities' livelihoods (Nawir et al., 2003).

In certain respects, the CCP program developed in Java is much more limiting and less flexible compared to that in the Outer Islands. Moreover, the content of the partnership contract suit *Perhutani*'s needs more than those of the people and community groups. The PHBM model commonly adopted in Java allows farmers to use allotments inside the company's concession to grow cash crops, fodder trees and fruit tress between the rows of teak trees. While the yield from cash crops, fodder and fruit tress remain largely with the farmers[9], some of the proceeds of selling the teak (25%) are distributed to the community through 'Forest Village Community Institution' (LMDH) established by *Perhutani* (Djayanti, 2006). The farmers can only use the allotment for up to two years until teak is expected to be free from weeds or any other competing plants and they

---

[8] This matter is clearly stipulated in all forestry laws starting from Forestry Law No.5/1967 to Government Regulation No.6/2007 regarding Forest Management (see Maturana et al., 2005and Hendra, 2007)

[9] In some cases, farmers might be required to share their cash crop selling profit with Perhutani

have to move on. However, there is no technical assistance provided by *Perhutani* in terms of institutional building and agricultural inputs (Awang et al., 2008).

In contrast, CCP programs implemented by timber concessionaires in the Outer Islands cover tree growing within concession areas that are claimed by forest dwellers, in surrounding concessions areas and/or on community owned lands. Companies cover all the costs for plantation, technical assistances and extension activities. Communities are required to plant 90 to 100% of the area with timber crops and the rest with they preferred cash crops. All yields from cash crops belong to the farmers, while profits from timber selling are shared between company and tree-growers in various proportions: 50:50, 60:40, 80:20 or 90:10. Companies retain the right to conduct harvesting, and communities hold the right to supervise jointly-managed lands and planted timbers. However the latter are obligated to ensure the maintenance of land ownership and the security of planted timber from theft and fires. In addition, contract duration ranges between 43 – 45 years, with thinning cycles in and between (see Nawir and Gumartini, 2003).

IV. Benefits and Constraints of the past strategies

4.1. Farm Forestry Program

4.1.1. Benefits

a. Flexible decision for land management

One of the benefits of farm forestry approach is that farmers have the ability to make decision about land management. Although particular credit schemes such as KUK-DAS (see 2.1.1) require farmers to plant particular tree species in their land, they do not restrict them from planting other 'cash crops'. The people, thus, have the opportunity to meet their various subsistence needs while at the same time benefit from the provision of resource support.

b. Provision of financial and resources support

The HR program is also beneficial as it provides funding and resource support (seedlings, technology and training) for the local communities. The provision of funding and related resources enables farmers to improve their land's productivity as well as enhancing the quality of the trees planted. In practice, however, farmers have to deal with intricate procedures to obtain this. For example, farmers may be eligible for 'farm forestry credits' (*Kredit Hutan Rakyat* / KHR) by grouping together until they reach a total planted area of 900 hectares, and find a business partner to administer the loan.

4.1.2. Constraints

a. Land and crop tenure security

Land and crop tenure security are often confused by the overlapping institutional arrangement of land ownership, particularly between customary/traditional arrangements and state regulations and laws. The basic laws, such as the *Agrarian Law 1960* and the *Forestry Act 1991*, only recognise customary/traditional arrangements half-heartedly. The *Forestry Act 1991* acknowledge customary land tenure arrangements only if they do not overlap the national interest and only as far as the use of above-ground resources. The agreements do not include below ground resources such as mining resources. The *Agrarian Law 1960* only approves a land certificate issued by the Agrarian Office as proof of land ownership but this has not been fully adopted by most traditional communities. As the result, customary/traditional arrangement is often undermined (McCarthy, 1999; Fay et al, 2000; Contreras, 2005). Prior to 1980, there was an alternative way to accommodate the existence of customary land tenure arrangements through the issuance of a notification letter by the village head. Yet this is now very weak under formal laws (Nawir and ComForLink, 2007).

b. Complex selling procedures and high transaction costs

Another constraint to the FF program is the strict regulation and complex procedures for selling log or primary wood products (e.g. square log or sawn timber) which come from private forest in Indonesia (Kusumanto and Sirait, 2000). In order to be able to market log or primary wood products from community-owned land, the farmers must purchase 'the Permit of Timber Utilization from Community-Owned Land' (*Izin Pemanfaatan Kayu dari Tanah Milik: IPKTM*) (as per decree No.20/Kpts-II/1997). This requires farmers to undergo intricate procedures starting from land ownership documentation verification[10] up to endorsement by the district government head / *Bupati* (Kusumanto and Sirait, 2000). Such intricate procedures open up the possibilities for rent-seeking by the government and intermediaries (broker). Farmers' also have to pay per-cubic levy of timber stock as well as other non-formal miscellaneous fees, such as costs for purchasing reference letter for logs or or primary wood products' form community-owned land (*Surat Angkutan Kayu* / SAK) (BAPPENAS, 1998). The costs accumulated from such a complex procedure are often too expensive to be covered by low-income farmers and often hinder them to participate in tree-growing activities.

<u>c. Poor market information and access</u>

The growing local and export-scale wood market has been one of the reasons behind the success of FF program (Nemoto, 2002; Nawir et al., 2008). Farmers, however, are still unable to get the most benefit of such condition and only get the small profit from log and/or processed wood production (Nawir et al., 2007). There are several reasons for such condition. Firstly, farmers have minimum market information and often unable to deal with complex log-selling procedure and high transaction costs; these then cause the farmer to sell their wood in lower prices (Nawir et al., 2007) (Nawir et al., 2007); Secondly, wood processing industries have an exclusive agreement[11] with the government which hinder them from purchasing wood from community-owned land (Byron, 2001).

<u>d. Low capacity of the farmers and minimum access to germplasms</u>

In most cases observed, small-scale tree growing activities are characterized by limited proactive management and planning. Gunasena and Roshetko (2000) argue that this is partly because farmers often lack technical capacity. It is also due to minimum institutional and capacity building facilitation provided by the local government, often caused by the lack of capacity in the local government. This, along with the problem of minimum access to 'high quality gerplasms (Gerhard et al., 2004), has often resulted in low quantity and quality products.

---

[10] Proof of land ownership can also be restraining as traditional communities often unfamiliar with formal land certificate system (Kusumanto and Sirait, 2000).

[11] Wood processing companies are required to submit an annual plan of sustainable supply (*Jaminan Pasokan Bahan Baku Berkelanjutan*) which says that industry is able to fulfill its installed capacity. This lead industries to seek for cooperation with plantation companies and/or forest concessionaires to secure their supply but at the expense of small-scale wood producers.

Enhancing smallholder management skills has been suggested in order to improve the productivity of small-scale tree growing system. Key skills include species selection/site matching; identifying tree farming systems that match farmers' land, labor and socioeconomic limitations, tree management options to produce high quality products; pest and disease management; and soil management (Gerhard, 2004).

4.2. Community Forestry (CF) Program

4.2.1. Benefits

a. Increasing access to forest land and forest resources

One of the obvious benefits of the CF program is the greater opportunity for people, particularly the local and forest-margin people, to become involved in the management of state forest areas (Awang, 2003; Subarudi et al., 2004). By engaging people, it is expected that the responsibilities for protecting the forestlands will be shared and the pressures over the forestlands, both natural and human-induced, can be minimized (Awang, 2003). But on top of all, that the livelihoods of the locals are expected to improve because of the greater access to various wood and non-wood forest products (Awang, 2003). Although the impact of CM, particularly on livelihood improvement, has been argued to be limited (for example, Kusumanto and Sirait (2000); Kusumanto, et al., (2005); Safitri (2006)), some still see it as an 'avenue' for the formal recognition of local rights and ownership over the forestland and forest resources (for example, Poffenberg (2006)) which would had been less possible prior to 1990.

4.2.2. Constraints

a. Land and resource Tenure

Kusumanto and Sirait (2000) argue that land and resource tenure status in the context of community forestry is somehow uncertain as to the rights and ownership over the forestland and the resources of the forestlands. Do the communities really hold the rights over the forest they established? Or does the state maintain full control over the forest lands and the resources on the lands? If so, what is the basis of the state's control? Would it be based on the inseparable connection between the forest lands and the forest resources or based on the principle of horizontal separation between forest land and forest resources? While the Indonesian *Agrarian Law 1960* recognizes the second principle, Forestry Act 1991 *i*s still uncertain. For example, forest areas allocated for commercial timber plantation (*Hutan Tanaman Industri* / HTI) are based on the principle of separation between the forestland and the forest resources, while Community Forestry program adopts the concept of an inseparable connection between the forestlands and the forest resources. Safitri (2006) argues that inconsistency found in the forestry law is owing to the government's preference to support elite corporations. In order to overcome the problems of resource tenure, benefit sharing has been recommended as one of the ways, yet this has not been taken into account appropriately in the implementation (as indicated in Nawir et al., 2007).

b. The weakness of local institutions

Strong local institutions, either in terms of the rules or the entity, are important to support the implementation of community forestry programs as they can stimulate collective action and accelerate the adoption of changes and opportunities (Byron, 2001; Molnar et al., 2004). Local institutions, however, have several weak points such as hierarchical decision making processes (Poffenberg, 2006), and a lack of conflict resolution and enforcement mechanisms (Le Vine, 1961). Hierarchical decision making can result in problems such as a lack of trust (Poffenberg, 2006) and lack of representation which then impede participation (Poffenberg, 2006). The lack of internal mechanisms for conflict resolution and enforcement make local institution depend on external parties such government and NGOs to solve their internal problems (Le Vine, 1961). Continuous facilitation for institutional and capacity building is suggested as one way to strengthen local institutions (Molnar et al., 2007).

c. Minimum technical assistance

The community forestry policies and regulations in Indonesia have been changing rapidly since its formal inception in 1995. Each of these policies has particular attributes which create complications in their implementation. This has become much more complex since the decentralization policy was put in place in 1999. As a result, stakeholders at local level exhibit different responses and reactions to these overarching laws and regulations (Safitri, 2006). Technical assistance is therefore essential to minimize such gaps in the field. Yet, this been argued as minimum on the ground (Kusumanto and Sirait, 2000; Suhardjito et al., 2000), due to the lack of incentives for capacity building (Awang, 2003) and the low capacity of the local government (WRM, 2004).

d. Complex licensing procedures

Kusumanto and Sirait (2001) describe that the complex licensing procedure has also been a significant constraint to the implementation of CF programs. In order to obtain a CF concession license, the community groups must through complex procedures, starting from village level up to ministerial level. Such long procedures are not only 'time-consuming' but also create high transaction costs and they often discourage people from participating in CF programs. With the decentralization system now in place, one may hope that such intricate procedures will now be simplified. Yet the central government is still holding on to its strong authority in terms of the designation of forest land as CF concessions. This indicates that the licensing procedure complexities are less likely to be resolved immediately time unless the central government is willing to cut down the bureaucracy and allow the district government exercise more authority.

e. Limited access to end products

The successful implementation of CF program has also been constrained by the short life of the concession. One of the reasons for farmers to participate in tree growing activities under the CF program is to benefit from the wood products of the trees they plant. In Indonesia, the trees are considered harvestable when they reach a diameter of about 30 cm (this is similar to 30 years of age; assuming annual increment of a tree is 1 cm/year), while the duration of CF concession is only 25 years. This implies that the people planting the trees will have no access to the harvested wood products as the concession will expire before the harvest cycle begins.

4.3. Community-Company Partnership Program

4.3.1. Benefits

a. Provision of job opportunities

Nawir and ComForLink (2007) argue that part of the benefits to be gained from the CCP program is increasing job opportunities for the local people. Through the CCP program, the company encourages the local people to be involved in various tree-growing schemes such as those located in the companies' domain which is occupied by the communities, in the vicinity of plantation concessions or within community-owned land (Nawir and Gumartini, 2003) for details). This not only provides the companies with reliable future sources of timber, but also provides job opportunities and extra income to local people from harvested timber at the end of each rotation under a long-term contract[12] (Nawir and Gumartini, 2003) which serves as a risk buffering and retirement fund as well (Nordwijk et al., 2007).

b. Financial and technical Support

Another benefit from the CCP program is the provision of financial supports such as social fund and credit assistance, and capacity building for local people such as intensive technical assistance for the participating households (Nawir et al., 2003; Nordwijk et al., 2007). This includes integrated pest management assistance for guarding against monoculture crop failure (Nordwijk et al., 2007) and the provision of 'superior' germplasm to the community (Maturana et al. 2003; Nawir et al. 2003; Nordwijk et al., 2007).

4.3.2. Constraints

a. Insecure land and crop tenure

The CCP program which is located in the community domain is generally secure in terms of land and resource ownership. The main problem usually comes from alternative CCP schemes located on community-owned land. Nawir and and ComForLink (2007) argue that it is still difficult for this particular type of program to be developed as legal frameworks with respect to land and resource tenure are still conflicting, as exemplified in the farm forestry program (see 3.1.2).

b. Potential loss of livelihoods

The CCP program, particularly if it is located on community-owned lands, only allows farmers to grow limited types of plants, with a strong preference for monoculture system (Noordwijk et all., 2007) and less flexibility for farmers in terms of land management (Nordwijk et al., 2007). Under such arrangements, the people lost the opportunity to allocate the land for other alternative uses. This then creates a potential livelihood loss for local people, particularly when unexpected hazards occur or in the case of market failure.

---

[12] Some of these schemes have reached harvesting cycle in the last 2005 in which the benefit sharing should have been allocated to the community (see Nawir and Gumartni, 2003)

c. Lack of local stakeholder involvement

Nawir and ComForLink (2007) argue that the CCP lacks key local stakeholders' participation. The community awareness campaigns and the design of community empowerment model have been carried out exclusively by commercial companies and the central government without the active participation of local NGOs and the government. This has resulted in weak local control over the contents of negotiations between the companies and communities and often ends up with companies being too dominant in directing the content (Nawir and ComForLInk, 2007). However, key local stakeholders are often not well prepared in providing feedback for better institutional arrangements (Nawir and ComForLink, 2007) and this is again generally due to the low capacity of the local stakeholders. Minimum involvement of local stakeholders then leads to other problems such as a lack of trust by the local community and failure to meet the commitments stipulated in contract agreement. According to my own personal experience, an active role by local stakeholders, particularly local NGOs, is imperative in building trust and directing the commitment of both negotiating parties.

d. Economic and financial challenges.

Nawir and ComForLink (2007) argue that another constraint for the CCP implementation is companies' ability to provide funding support in the long term and acquiring little or no collateral loan facilities from commercial banks to develop partnerships. Other related issues include the availability of markets with more favorable prices and improving the quality of timber produced from community- owned land which often does not meet market standard.

V. People Plantation Forest Policy

The main concerns of the 'People Plantation Forest' policy are to enhance the forest sector's contribution to alleviating poverty (pro-poor), provide greater job opportunities (pro-job) and improve the quality of development through proportional investment among economic actors (pro-growth) (Ministry of Forestry, 2006).

The management aspect of People Plantation Forest (PPF) concession rights (IUPHHK-HTR) is regulated specifically in The Ministry of Forestry regulation (*Permenhut*) No. 23/2007) as well as supported by government regulation (PP)

No.6/2007. Details of the management aspect of People Plantation Forest are summarized in Table 3.

The People Plantation Forest policy has some specific attributes maintained as a result of past strategies (Nordwijk et al., 2007). Some of these attributes: a longer management time frame; the establishment of BP2H (Badan Layanan Umum Pembiayan Pembangunan Hutan), a service unit specifically assigned to deal with financial assistance for the development of PPF; the establishment of village financial units (Lembaga Keuangan Pedesaan) aimed specifically at supporting the poorer members of the community who cannot access credit services or other formal funding schemes toward the development of PPF at the village level. Furthermore, the mechanism for PPF development is designed flexibly to suit various contexts.

Box 1.	Three Alternative Models stipulated in Ministry of Forestry Regulation No. 23/2007 (Permenhut) regarding People Plantation Forest (PPF) Policy

The three PPF models proposed as included in the Permenhut:

<u>1) The PPF independent model</u>
The independent PPF schemes are implemented directly by local communities, which are expected to establish farmer groups and submit proposals to the district heads. Subsequently, on the recommendation of the district head, the government allocates areas and IUPHHK-HTR (PPF Concession) permits to the individual members of these groups.

The head of each group is responsible for the development of the PPF, applications for and repayment of loans, as well as organising markets for any timber produced. In addition, every member is responsible for reminding the other members to fulfil their obligations. Local governments will assist these independent schemes to develop PPF.

<u>2) The PPF partnership model involving state-owned or private HTI (Timber Plantation) companies</u>
In this HTR development model, local communities establish groups, which the district head then recommends to the Ministry of Forestry. The government grants IUPHHK-HTR permits to individuals and determines a partner for each group. The appointed partners are responsible for inputs, training, assistance and making markets available for any timber produced.

<u>3) The PPF developer model</u>
In HTR development through the developer scheme, private or state-owned companies act as planters. The government then surrenders the plantation forests to designated communities (individuals/groups) willing to take on the responsibility of becoming developer scheme IUPHHK-HTR holders. Any expenses for developing plantations are calculated as loans to IUPHHK-HTR holders to be paid back in stages in accordance with the credit agreements.

Source: Nawir and ComForLink, 2007

Management aspect under IUPHHK-HTR	Conceptual management of HTR
1. Recipients	Individuals, and/or cooperative
2. Management time frame	The longest is 60 years based of Permenhut (100 years based on PP 6/2007)
3. Area	The maximum is 15 ha per family applicant, or based on the enterprise capacity of the cooperative
4. Main species	Woody forestry species or a combination of different species (e.g. woody forestry species combined with annual horticultural species)
5. Funding	Funded through the service unit initiated specifically for forestry development or Pola Pengelolaan Keuangan Badan Layanan Umum Pembiayaan Pembangunan Hutan (BP2H)
6. Price set-up	Prices used refer to the set base price (harga dasar) defined by the MoF
7. Endorsement of the right	Bupati (District Head) on behalf of the MoF
8. Responsible agency for improving technical development	Directorate General of Forestry Production Management (Ditjen Bina Produksi Kehutanan

	- BPK)
9. Supervision and monitoring	• Head of Provincial Forestry Services and District Forestry Services to supervise and control the progress of HTR • Head of BP2H to control and evaluate the use of the loans • Head of the village to supervise the implementation of the HTR development
10. Conditions for cancelling the rights	• Transferring the rights without written approval from the granting authority • No activity for one year • Fails to implement the silviculture practice suitable to the local conditions • No long-term working plans for timber utilization (Rencana Kerja Usaha Pemanfaatan Hasil Hutan Kayu - RKUPHHK) is prepared within one year of the permit being issued • Fails to put together the annual working plan (Rencana Karya Tahunan – RKT) in two months at the latest • Leaves the areas • Criminal proceedings are applied • Passed away.
11. Sanctions (similar to HTI company)	• Ten times that of the set base price for any violation in relation to: submitting RKT in two months at the latest before the working RKT; measurement/testing the forest production; and developing the long term RKUPHHK • Fifteen times that of the PSDH for any violations in relation to: timber cutting to make the corridor before any permit is granted or does not adhere to the permit granted.

Adapted from: Nawir and ComForLink (2007:11)

Table 3. Management aspect under 'People Plantation Forest' policy

VI. DISCUSSION

This section provides an analysis which focuses on whether or not The People Plantation Policy, particularly the independent model, will be able to stimulate tree growing activities in Indonesia by minimising and/or removing key impediments and disincentives. The key impediments used as variables for the analysis are adapted from Byron (2001) and the key themes are those that emerged from the experiences of the existing strategies (as discussed in section 3 of this paper). The Illustrations of some of the key impediments and disincentives to be discussed is presented in Figure 3.

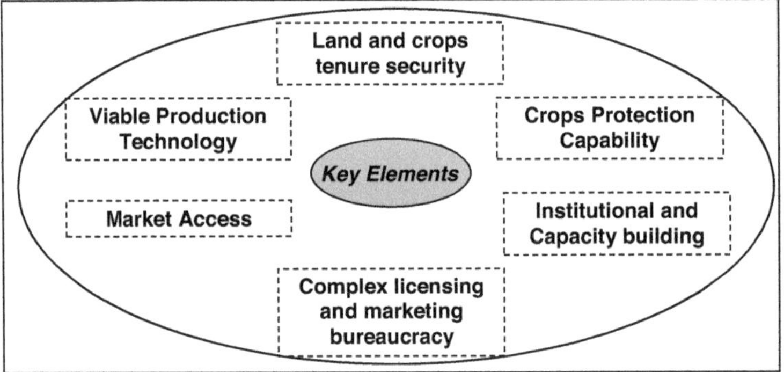

Figure 1. Key elements (impediments) used for analysis used for analysis
Source : Adapted from Byron (2001) and Nordwijk et al.(2007)

6.1. Land and Crop Tenure Security

Tenure security can be associated with both legal construction and state of mind. Place et al. (1994) define tenure security as an individual perception of having a piece of land or resources on a continuous basis, free from imposition or interference from outside sources and gaining the benefits of labour and capital invested in that land, either in use or upon transfer to another holder. Lindsay (1998) suggests that the extent of tenure security can be measured by using several indicators: the clarity of the rights; legal certainty; appropriate duration; enforcebility; exclusivity; the clear legal status of right holders; and the proper position and authorities of the government in granting the rights.

In general, most of the elements of tenure security have been well represented in the Ministry of Forestry Regulation (*Permenhut*) No.23/2007. However, there is a question with respect to the transferability of the rights. *Permenhut* has clearly stipulated that once issued, the license will be valid for 60 years duration but will be revoked if the holder passes away (See Box 1). This latter provision has created ambiguity particularly regarding the security of resources already invested in the land and the financial liabilities of the investment (there is who will these be passing on to?). In this regard the transferability of tenure is particularly important to ensure the sustainbility of PPF development. Furthermore, tenure security has also been linked with external factors such as unresolved problems of overlapping of authority between central and local governments, and inter-sectoral regulations.

6.2. Viable Production Technology

The success or failure of tree planting activities is determined by a variety of factors. One of them is the biological factor which includes seedling quality, more effective fertiliser treatment, innocullation with microsymbiants, and superior germplasm (Cromwell et al., 1992; Byron, 2001). Such factors are particularly important to increase land productivity per hectare, as well as reducing costs or the risk of disease or pest outbreak. *Permenhut* recognises the importance of such technology as far as the provision of production inputs (such as seedling, fertiliser and pesticides), yet does not specify in detail as where to source those inputs from. What quality of inputs are to be used? How to treate and/or manage those inputs? These questions appear to indicate that biological factor has not been seriously taken into account.

Supply of high quality seeds or seedlings has been argued as one component which is still difficult to achieve (Nordwijk et al., 2007), particularly for local farmers. This is because supply of superior tree seedlings in Indonesia is still conrolled by the formal sector such as specific government agencies (often under central level coordination), research organisations and forest industries (Harwood et al. 1999 cited in Nordwijk et al. 2007). Therefore, it is important to create the link between local farmers and these institutions to tackle the problem. Otherwise, the role and involvement of NGO, which is one of the main feature of the PPF policy, should be directed toward creating networks between farmers and seed providers.

6.3. Market Structure and Access

Insecure market access has been the major impediment for the success of small-scale tree growing activities. Under the independent model of PPF, such a problem is likely to be perpetuated, because, for one thing the main responsibility for market establishment lies in the hands of farmers themselves. Technical facilitaton which is stipulated in *Permenhut* still unclear. *Permenhut* does not clarify the mechanisms to be used for facilitation, for the institutions to be in charge for the incentives to be provided for marketing facilitation in the independent model. This is in contrast to the partnership and developer model in which market facilitation is handed over to industries and contractors. What is likely to occur is that the facilitation will be conducted either by the central government or local stakeholders. Each of these options has its own advantages and disadvantages. Central government may provide various physical and financial incentives but often lacks the capacity to encourage community participation, partly due to the lack of trust and perhaps intensive facilitation. With the local stakeholders in charge, it will be the other way around. Thus, it is perhaps better for the central government and local stakeholders to share the responsibility for market assistance.

6.4. Crops Protection Capability

The duration of a PPF concession ranges from 60 – 100 years (Table 3). With such a long time frame, farmers need to ensure that their investment will 'survive to marketable maturity' (Byron, 2001:223); this includes protection against potential risks which come from natural or human-induced factors. Despite the significance of long-term crop protection, neither *Permenhut* nor government regulation (PP 6/2007) suggests any mechanisms as such. It is also not clear whether this aspect will be factored in to annual working plan (RKT) or long-term working plan for timber utilisation (RKUPPHK) (see Table 3 for details) which farmers must provide.

6.5. Local Institutional and Capacity Building

The importance of local institutions and arrangements in encouraging community participation in small-scale tree growing activities has been discussed in the literatures (for example Awang et al., 2004). This particular aspect is strongly embodied in *Permenhut* . For instance, *Permenhut* prioritises 'community cooperatives' which are 'established by the communities living in or surrounding forest areas' for the recipients of the PPF program. This emphasises the importance of locally-established institutions, instead of government-extablished institutions which characterise past government intervention. Furthermore, *Permenhut* also mandates key stakeholders at the local level such as the district government head (Bupati), village head and local NGOs to cooperate in facilitating institutional and capacity building, as well as providing financial and technical assistance in this regard. Therefore, the PPF frawework appears to cope well with this issue.

6.6. Complex Licensing and Marketing Bureaucracy

From *Permenhut* (see Table 3) and the regulation (PP 6/2007) perspective, it appears that the PPF policy does not cope well with the problem of licensing bureaucray as rights endorsement remains The Ministry of Forestry's and hence farmers or community groups still have to deal with long and intricate procedures. Nevertheless, the latest revision of *Permenhut,* The Ministry of Forestry Regulation No.8/2008 has indicated positive feedback as now the MoF's authority goes only as far as indicating the designated area while the district head (Bupati) holds the authority to issue the rights on behalf of the MoF.

Marketing bureaucracy can be divided into two main processes: pre and post logging. Pre-logging processes range from the verification of land ownership up to thw issuance of the IPKTM permit (see 3.1.2.b); Post-logging processes cover the process of transporting timber from log-yard to market. Regulation can only deal with the post-logging process, while pre-logging proces is within local government's authority zone. Hence, simplification of marketing bureaucracy will be highly dependent on the local government's commitment. Once more, vertical coordination is required in this regard.

VII. Conclusion

The Indonesian government has been experimenting with small-scale tree growing activities over the last three decades. During this period, there have been a number of strategies put in place which can be categorised into three main approaches: farm forestry (*Hutan Rakyat*); community forestry (*Hutan Kemasyarakatan*) and Company-Community Partnerships (*Kemitraan Perusahaan dan Masyarakat*). Each of these strategies has its own benefits and limitations; however, the impact on the ground has been largely viewed as limited. This is partly because of the small percentage of forestland which is actually under community management and the lack of improvement of livelihoods of the people living in and near forest areas.

In 2007, the government issued the 'People Plantation Forest' policy which brings with it specific attributes which is cumulative results of past strategies (Nordwijk et al., 2007). Amongst these attributes are a longer management time frame, the establishment of specific task forces to assist in field implementation, the active involvement of local key stakeholders and flexible investment models.

Despite the new breakthroughs, questions still arise about the extent to which this program can really eliminate the key obstacles to its success in order to stimulate independent tree-growing activities. The evidence so far indicates that the PPF policy might cope well with the issue of local institutional and capacity building, and to some extent, with issues such as viable production technology and market access. However, the PPF policy does not yet seem alike to deal with major constraints such as land and crop tenure security, and the overly complex licensing and marketing bureaucracy. These arise from the existing problems of overlapping inter-sectoral regulations and jurisdictional fragmentation between central and district governments. Hence, it is less likely that the PPF, as it stands, will be able to push forward independent tree-growing activities.

Although there is no single solution to this situation, increasing and continuous coordination between governments across different jurisdiction (i.e. forestry, agrarian and planning and development agency) as well as along vertical hierarchy by taking seriously into accounts local stakeholders view is still an important issue towards minimising never ending constraints such as land and crop tenure security. Governments at different jurisdiction and level should also support participatory mapping processes which have been underway in some areas by some local NGOs to map out traditional land use type as this. In many cases, is viewed as legitimate source of land demarcation by the local in which by having governments endorse it would obviously be an incentive for the local people to invest in tree growing, assuming that other incentives are also provided. Past experiences had also indicated that increasing coordination between governments along vertical hierarchies and with local stakeholders often result in synergizing efforts in establishing and strengthening local institutions and communities through the provision of planting subsidies, technical and financial supports, as well as access to market. Close coordination to synergise commitments and supports in minimising major constraints, however, should be maintained at least to the point when the implementation of PPF is close to its level of sustainability.

REFERENCE

Agar, M.H., 1983. Political Talk: Thematic Analysis of a Politic Argument, Review of Policy Research, 2(4):601-614

Awang, S.A, 2003. Politik Kehutanan Masyarakat, Center for Critical Social Studies bekerjasama dengan Kreasi Wacana Yogyakarta.

Awang, S.A., Widayati, W.T, Himmah, B, Astuti, A., and Wardhana, W., 2004. Collective action on state forest company in Java, Indonesia. Source available from: http://dlc.dlib.indiana.edu/archive/00001832/00/Awang_SA.pdf.

Awang, S.A.; Widayanti, W.T.; Himmah, B.; Astuti, A.; Septiana, R.M.; Solehudin; Novenanto, A., 2008. Panduan Pemberdayaan Lembaga Masyarakat Desa Hutan. Source available from: www.cifor.cgiar.org/Publications/Detail.htm?&pid=2457&pf=1-7k

Boyatzis, R.E. 1998: Transforming qualitative information: thematic analysis and code development. Sage.

Braun, Virginia and Clarke, Victoria(2006) 'Using thematic analysis in psychology', Qualitative Research in Psychology, 3(2): 77 — 101

Byron, R.N., 2001. Chapter 16. Keys to Smallholder Forestry in the Tropics. *In* Harrison, S.R., and Herbohn, J.L. (Eds), *Sustainable Farm Forestry in the Tropics: Social and Economic Analysis and Policy*. p.195-210.

Contreras-Hermosilla A., Fay C. 2005 Strengthening Forest Management in Indonesia through Land Tenure Reform: Issues and Framework for Action. Washington, DC. and Bogor, Indonesia Forest Trends and World Agroforestry Center. 2005. Source available from: http://www.forest-trends.org/documents/publications/Indonesia%20Report_final%208-22-05.pdf

Cromwell, E, E. Friis-Hansen and M. Turner. 1992 . The Seed Sector in Developing Countries: a Framework for Performance Analysis. Overseas Development Institute, Working Paper No. 64. 71 p.

Daniel, J, Verbist, B., Carandang, W.M. , Kaomein, M., Mangaoang, E., Nichols, M., Pasaribu, H. and Zeiger, Z. 1999. 'Working Group 4 – Linkages for training and information dissemination'. In J.M. Roshetko and D.O. Evans (eds), Domestication of agroforestry trees in Southeast Asia. Forest, Farm, and Community Tree Research Reports, special issue 1999, pp 226-228.

Djayanti, D., 2006. Managing forest with Community (PHBM) in Central Java: Promoting Equity in Access to NTFPs, (Edited by: Sango Mahanty, Jefferson Fox, Leslie Mclees, Michael Nurse, Peter Stephen), Perun Perhutani, p.63-82.

Fay, Chip., and de Foresta, H., 1998. Progress towards Increasing the Role Local Play in Forestlands Management in Indonesia. ICRAF Bogor Indonesia. Source available from: www.worldagroforestry.org/sea/Publications/files/workingpaper/WP0045-04.PDF

Fay, C., Sirait, M. and Kusworo, A., 2000. Getting the Boundaries Right: Indonesia's Urgent Need to Redefine its Forests Estate, World Agroforestry Center, Bogor, Indonesia

Fay, Sirait and Kusworo. 2000. Quoted in Colchester, 2002. 'Bridging the Gap: Challenges to Community Forestry in Indonesia'.

Gunasena, H.P.M. and Roshetko, J.M. 2000. Tree Domestication in Southeast Asia: Results of a Regional Study on Institutional Capacity, International Centre for Research in Agroforestry (ICRAF) Bogor, Indonesia. 86 pp

Hindra, B., 2005. Indonesian Community Forestry 2005. Source available from: http://www.dephut.go.id/INFORMASI/RRL/RLPS/Bangkok.pdf

Indonesian National Planning and Development Agency (BAPPENAS), 1998. Legislation and Policy for Fire Prevention and Drought Management, Working Paper No.3 - Asian Development Bank TA 2999-INO. Source available from: http://www.adb.org/Documents/Reports/Fire_Prevention_Drought_Mgt/default.asp -19k

Joffe, H and Yardley, L., 2004. Chapter 4. Content and Thematic Analysis. In Research Methods for Clinical and Health Physicology, P.56-58

Kusumanto, Y., and Sirait, M.T., 2000. Community participation in forest resource management in Indonesia: policies, practices, constraints and opportunity. Source available from: http://www.worldagroforestry.org/downloads/publications/PDFS/http://www.world agroforestry.org/sea/Publications/files/workingpaper/WP0046-04.PDF

Kusumanto, T., E.L. Yuliani, P. Macoun, Y. Indriatmoko and H. Adnan. 2005. Learning To Adapt: Managing Forests Together in Indonesia. CIFOR, YGB and PSHK-ODA. Bogor, Indonesia

Le Vine, R.A., 1961. Anthropology and the Study of Conflict: An Introduction. The Journal of Conflict Resolution, 5(1):3-15.

Lindsay, J.M., 1998. Designing Legal Space for Community-Based Management. Paper presented at International Workshop on Community-Based Natural Resource Management. Washington, D.C.: The World Bank, May, 10-14.

Manurung, G.E.S., Roehetko, J.M., and Budidarsono, S., 2004. Traditional tree farming system in West Java and their importance to local people. Source available from: http://www.worldagroforestry.org/sea/products/training/GroupTra/SAFODS/data/ Abstract/Theme1_Tradisional%20Tree%20Farming.pdf (access date: 24 April 2008).

Maturana, J., Hosgood, N and Suhartanto, A.A., 2005. Towards Company-Community Partnership, CIFOR Working Paper No.29(i). Source available from: http://www.cifor.cgiar.org/mla/download/publication/partnership.pdf

McCarthy, J., 1999. Village and state regimes on Sumatra's forest frontier: a case from the Leuser ecosystem, South Aceh. Resource Management in Asia-Pacific, (RMAP) Program, RSPAS. ANU, Working Paper No.26, Source available from: http://hdl.handle.net/1885/40980

Meine van Noordwijk, S Suyanto, Suseno Budidarsono, Niken Sakuntaladewi, James M. Roshetko, Hesti L. Tata, Gamma Galudra and Chip Fay, 2007. Is Hutan Tanaman Rakyat a new paradigm in community based tree planting in Indonesia?, ICRAF working paper no.45.

Michon, G. 2005. Domesticating Forests. How farmers manage forest resources. Center for International Forestry Research and World Agroforestry Centre. 187 p.

Ministry of Forestry, 2006. Peraturan Menteri Kehutanan No.23/Menhut-II/2007 tentang Tata Cara Permohonan Izin Usaha Pemanfaatan Hasil Hutan Kayu pada Hutan Tanaman Rakyat dalam Hutan Tanaman (Procedure for Utilisation Permit of Wood Forest Product from People Plantation Forest).

Molnar et al., 2007. Community-based enterprises: their status and potential in tropical countries. ITTO Technical Series #28. Source available from: http://www.itto.or.jp/live/Live_Server/3710/ITTO_TS28.pdf

Nawir, 2008. Strategies to enhance the Implementation of small-scale commercial tree growing in Indonesia. Working Thesis, 1 February 2008.

Nawir, A. A. and L. Santoso. 2005. "Mutually beneficial company-community partnerships in plantation development: emerging lessons from Indonesia, International Forestry Review,7(3): 177-192

Nawir, A.A. & ComForLink, 2007. Forestry companies' perspectives: Improving community roles in plantation forest development through partnerships. Paper for the World Bank as a contribution to the national meeting organised by Dewan Kehutanan Nasional (DKN) - National Forestry Committee on: 'Konsolidasi dan Percepatan Pelaksanaan Restrukturisasi Kehutanan' (Consolidation and revitalisation in restructuring forest policy), Jakarta, 2-3 May 2007.

Nawir, A.A., and Gumartini, T., 2003. Company--community partnership outgrower schemes in forestry plantations in Indonesia: an alternative to conventional rehabilitation programmes. FAO Regional Office for Asia and the Pacific, Bangkok, Thailand.

Nawir, A.A., Hakim, M.R.H., Julmansyah, Ahyar, H.M.A. and Trison., 2007. Feasibility of Community-Based Forest Management Partnership with a Forestry District Agency (Case Studies: Sumbawa and Bima, West Nusa Tenggara), In Integrated Rural Development in East Nusa Tenggara, Indonesia. Proceeding of a workshop to identify sustainable rural livelihoods, held in Kupang, Indonesia, 5-7 April 2006. ACIAR Proceeding No.126.

Nawir, A.A., Kassa, H., Sandewall, M., Dore,D., Campbell, B., Ohlsson B., and Bekele, M., 2007. Stimulating smallholder tree planting – lessons from Africa and Asia, Unasylva (FAO), 58(228):53-58

Nawir, A.A., Murniati, Rombo, L., 2008. Forest Rehabilitation in Indonesia: where to after more than three decades?, Center for International Forestry Research Center Publication, Bogor, Indonesia.

Nemoto, A., 2002. Farm Tree Planting and the Wood Industry in Indonesia : a Study of Falcataria Plantations and the Falcataria Product Market in Java. Source available from:http://enviroscope.iges.or.jp/modules/envirolib/upload/371/attach/04_Indonesia.pdf.

Place, F., M. Roth., and P. Hazel., 1994. Land Tenure Security and Agricultural Performance in Africa: Overview of Research Methodology. *In* Bruce, J.W., and S.E. Migot-Adholla (Eds), *Searching for Land Tenure in Africa,* Dubuque: Kendal and Hunt Publishing Company.

Poffenberg, 2006. People in the forest: community forestry experiences from Southeast Asia., Int. J. *Environment and Sustainable Development*, 5(1):57-69

Pradhan, M., A. Suryahadi, S. Sumarto, L. Pritchett. 2000. Measurements of Poverty in Indonesia: 1996, 1999, and Beyond. Social Monitoring and Early Response Unit Working Paper June 2000.

Safitri, M.A., 2006. Change Without Reform? Community Forestry in Decentralizing Indonesia, Paper Presented at the 11th IASCP Conference, Bali, 19-23 June 2006. Source available from: http://www.indiana.edu/~iascp/bali/papers/Safitri_Myrna_Change.pdf

Subarudi, Idris, M.M., Achmad,B., and Iman, M.N., 2004. Community forestry for poverty reduction-lessons learned in Indonesia. FAO Publication. Source available from: http://www.fao.org/docrep/007/ad511e/ad511e0e.htm.

Suhardjito, D., Azis, K., Wibowo, D. Sirait, M., and Santi, E., 2000. Karakteristik Pengelolaan Hutan Berbasis Masyarakat (Characteristics of Community-Based Forest Management), Pustaka Kehutanan Masyarakat – Studi Kolaboratif FKKM.

- Editor's comment1   : What's the findings?

  Our comment   : Findings and suggestion have been included in the abstract-

- Editor's Comment2   : In Perhutani Unit III (Malang), the local community has to share some of their profit to Perhutani

  Our Comment   : Thanks for the input. we've adjusted the wordings from "remain" to "remain largely" and put the explanation in footnote

- Editor's comment3   : Constraint

  Our comment   : Agree. We have included sections mentioned in the constraint section d (minimum institutional and capacity building and business facilitation provided by the local government are causes for constraint explained in section d)

- Editor's comment4   : or Forestry Act 1991

  Our comment   : Agree, thanks. We have change the term to make it more consistent.

- Editor's comment5   : Please check the reference for Djamhuri, 2008, Place et al, 1994, Nawir et al, 2003, WRM, 2004, Kuswanto dan Sirait, 2000, Sumarto et al, 2003

  Our comment   : I have included Place et al, 1994, Nawir et al, 2003 and Kuswanto dan Sirait, 2000. I deleted Djamhuri, 2008 as readers can still find relevant examples from Awang et al., 2004. I deleted Sunarto et al., 2003 as relevant examples can still be found in Poffenberg, 2006. I changed WRM, 2004 to Le Vine, R.A., 1961 to make it more scholarly relevant.

Overall, we appreciate your valuable insertion and comments.

YOUR KNOWLEDGE HAS VALUE

- We will publish your bachelor's and
  master's thesis, essays and papers

- Your own eBook and book -
  sold worldwide in all relevant shops

- Earn money with each sale

Upload your text at www.GRIN.com
and publish for free